volume 4 DEVELOPMENT AND UNDERDEVELOPMENT 1945–1975	volume 5 THE GLOBAL COMMUNITY 1975–2000	volume 6 INTO THE 21ST CENTURY 2000–	chapter	topic
THE END OF PEASANT CIVILIZATION IN THE WESTERN WORLD	INDUSTRIALIZED AND MULTINATIONAL AGRICULTURE	THE PROBLEM OF SURVIVAL AND THE PROMISE OF TECHNOLOGY	1	Land, agriculture, and nutrition
OLD DISEASES IN THE THIRD WORLD AND NEW ACHIEVEMENTS IN MEDICINE	OVERPOPULATION, DEMOGRAPHIC DECLINE, AND NEW DISEASES	POPULATION, MEDICINE, AND ENVIRONMENT: A RATIONAL UTOPIA	2	Hygiene, medicine, and population
THE AGE OF PRIVACY: HOUSING, CONSUMER GOODS, COMFORT	MEGALOPOLISES IN THE THIRD WORLD; MULTIETHNICITY IN THE WEST	TOO MUCH AND TOO LITTLE: THE MISERY OF WEALTH	3	Living: environment and conditions
WHITE-COLLAR WORKERS, MANAGERS, AND LABOR PROTEST	AUTOMATION AND DECENTRALIZATION IN THE POST-FORD ERA	"THE END OF WORK" AND THE NEW SLAVERIES	4	Labor and production
NUCLEAR ENERGY: THE GREAT FEAR, THE UNCERTAIN HOPE	THE SEARCH FOR ALTERNATIVE ENERGY SOURCES	NEW FRONTIERS IN ENERGY	5	Raw materials and energy
THE PRODUCTION OF THE AUTOMOBILE	ELECTRONICS AND INFORMATION SCIENCE	WORK WITHOUT WALLS	6	Working: environment and conditions
MAN AND GOODS ON FOUR WHEELS	THE AIRPLANE IN MASS SOCIETY	FROM THE EARTH TO THE COSMOS: THE EXPLORATION OF SPACE	7	Transportation
EVERYDAY ENCHANTMENT: THE TELEVISION	THE INFORMATION AGE: COMPUTERS AND CELL PHONES	CYBERSPACE: THE WEB OF WEBS	8	Communication
PARALLEL ROADS: DEVELOPMENT AND UNDERDEVELOPMENT	THE COLLAPSE OF SOCIALISM AND THE RISE OF NEO-CONSERVATISM	THE MANY FACES OF GLOBALIZATION: LOCAL WARFARE AND THE GLOBAL COMMUNITY	9	Economics and politics
MOVEMENTS OF LIBERATION AND PROTEST: THE THIRD WORLD AND THE WEST	FEMINISM, ENVIRONMENTALISM, AND THE CULTURE OF UNIQUENESS	UNIVERSALISM AND FUNDAMENTALISM: THE NEW ABSOLUTISM	10	Social and political movements
CONSUMERISM AND CRITICISM OF THE CONSUMER SOCIETY	THE INDIVIDUAL AND THE COLLECTIVE	AFTER THE MODERN: ENVIRONMENTALISM, PACIFISM, AND BIOETHICS	11	Attitudes and cultures

THE ROAD TO GLOBALIZATION
Technology and Society Since 1800

In private and in public, at work or at play, in every stage of life, we live with technology. It becomes ever more present, and our perception of its artificiality fades through daily use. Within a very short time of their emergence, new possibilities seem to have been with us always, and the new almost immediately becomes indispensable. The choices that technology dictates and the paths that these choices take appear to be the only choices and paths possible—undeniable, unquestionable—and we perceive as natural the constructed world in which we live.

Despite the opportunities that technology affords us, and the promises that it makes constantly, we greet it with a general discomfort, an uneasiness that often does not reach the conscious level. But the manifestations of environmental crises can no longer be considered in isolation. The Westernization of the world marches in step with the widening—and already yawning—chasm between north and south, as well as with the emergence of aggressive localism. War seems to have resumed its role as a common tool in international confrontation. New diseases alarmingly outpace scientific discoveries, and biotechnologies and genetic experiments obscure the line between the human and the inhuman. The importance of the question of meaning has not been lessened by the decline of the sacred; but this question seems to find no place in the universal logic of growth that overcomes difference to guide governing bodies as representatives of economic and financial power.

A renewed uncritical faith in Progress on one hand and a demonization of "techno-science" on the other are often associated with a lack of context that comes of technology and with the dominance of a logic that neglects history. This logic can verify correctness in predetermined ways, but it does not comprehend the complexity of the greater process of change: it appreciates the present and the immediate future, but it cannot perceive itself as part of a larger historical evolution.

The first aim of a social and cultural history of the technology of the last two centuries, then, is to offer a careful and coherent study of the roads that have led to the development of modern culture. The basic objective of this series is the reconsideration of the innovative changes that have taken place and their diffusion over time, rather than a description of their first appearances. These innovative changes have marked and continue to determine our daily lives, the way we work, our relationships, and the points of view that contribute to global diversity.

It is important to recognize that our interpretations of the 19th and 20th centuries are centered on the men and women of the West, on their histories and cultures. This is undoubtedly a biased point of view, and it would be misguided to think that this partiality could be overcome by a simple updating of knowledge. The changing of a point of view that is rooted in history probably requires insight into processes that operate well beyond our perception. Perhaps the globalization that is underway, with its various worldwide effects, is establishing itself through precisely this mechanism: it is forcing a confrontation among lifestyles and different cultural models in new, absolute terms.

In general, the common historiography treats technological innovations only in brief digressions, glossaries, or chronologies of inventions and inventors; but we cannot fill in its gaps by constructing a separate history. Our realization of the economic, social, and cultural importance of industrialization, and our perception of the process as uninterrupted and ever more pervasive, have caused us to re-evaluate both the transformation itself and the new landscapes that industry has created—linking technology to economics and to politics, and systems of labor and production to culture and to social movements.

We can group as the Age of Technology the events that have been paving the road to the future for the last two centuries. Understanding the risks and the opportunities involved in so rapid a transformation of our world will require a change of mind and an updating of our culture—both of which are impossible without a broadening of knowledge and a renewal of historical consciousness.

4
DEVELOPMENT AND UNDERDEVELOPMENT
1945–1975

PIER PAOLO POGGIO
AND
CARLO SIMONI

ILLUSTRATED BY GIORGIO BACCHIN

IMPERIAL PUBLIC LIBRARY
P. O. BOX 307
IMPERIAL, TEXAS 79743

LAND, AGRICULTURE, AND NUTRITION

1. THE END OF PEASANT CIVILIZATION IN THE WESTERN WORLD

After World War II, agriculture in Western countries advanced very rapidly. In the Third World and in China, agriculture was at the center of widespread reform movements, with mixed results. These movements ultimately followed the Soviet model, which was fundamentally anti-peasant and, despite amazing promises made by Nikita Khrushchev (1894–1971), eventually bogged down in compromise among cooperatives, governmental agencies, and small landowners. The end of peasant civilization in the Western world was near.

North America took a completely different direction. Even with a decrease in the number of agricultural workers in the 1960s to 7% of the population, it easily satisfied its domestic demand and could export nearly 15% of its production. Europe followed the American model with slight delays, and within a few decades it saw its own defining revolution. The traditional peasant culture disappeared; the areas of cultivation decreased in size, and the countryside emptied out. In Italy, these occurrences were significant; in 1951, individuals still actively employed in the agricultural sector numbered nearly 50% of the population, but forty years later the percentage had decreased to 8.4%.

Throughout the West, excepting areas that specialized in fruit and vegetable farming, the land was ruled by machines, with only a slight human presence. Production increased spectacularly in both quantity and pace. In the 1950s, 30 labor-hours were required in the European farmlands to reap one ton of wheat; at the dawn of the 1980s, the work took 30 minutes. These successes were gained at a high price, though—increases in abandoned land, the use of chemical fertilizers and pesticides, pollution, and energy-consuming types of agriculture and husbandry.

The easy availability of inexpensive food, a situation that had not been seen before in the history of most of the world, led to a loss of interest in the rural world, causing its inhabitants to adopt an urban lifestyle. Peasants either found supplemental jobs, and cultivated their fields only part of the time, or became agricultural entrepreneurs. Industrial farming was the result of a system that centered on multinational agricultural, chemical, and pharmaceutical industries; these had technological and scientific laboratories and could influence the politics not only of nations, but also of international organizations like the European Community.

The agricultural politics of the European Community, then, which was

1. *High-altitude plowing, Merano, Italy, 1954. Although economically unrewarding, agriculture in mountainous regions and on inland hillsides has traditionally played a key role in safeguarding the territory. (Photograph by Elio Ciol.)*
2. *An abandoned village in the southern part of the Apennine Mountains in Italy. In the 1950s, over 10 million emigrants moved from southern to northern Italy. This biblical exodus left entire inland towns without inhabitants. (Photograph by Mario Giacomelli.)*
3. *Cassari, a village near Catanzaro, Italy, 1978. The hydrogeological shortfall was another cause of the abandonment of areas of the Apennines. The inhabitants of the nearby village of Ragonà, which had been declared uninhabitable, were assigned to these squalid houses, which were rapidly constructed to create a "re-formed village," as these southern Italian towns that rose from nothing are called. (Photograph by Francesco Faeta.)*
4. *A gigantic harvester/thresher. With the introduction of large rubber tires, increasingly complex agricultural machines, which could do the work of hundreds of people, appeared in fields around the world, creating another area of competition between the USA and the USSR.*

5. *A corn harvest in the United States. After its introduction in World War II, hybrid corn grown from crossbred seeds became a global phenomenon, centered in the USA. The increased production was used mainly to feed cattle. (Photograph by Burt Glinn/Magnum/Contrasto.)*

characterized by complicated rules, can be seen as an attempt to guide a very rapid historical transition—painlessly, and with huge concessions to the most influential producers—and to avoid disruptive social and political effects.

The cost of this was great waste, with frequent switching between free trade and protectionism in a vain attempt to reconcile the claims of farmers, consumers, and multinational enterprises. Particularly harmful was Europe's delay in recognizing the importance of particular regions and localities in achieving an agricultural output of high quality. The choice of maximum productivity, based on the industrial model of efficiency and profit, began to result in serious problems in social life, the environment, and territorial security.

HYGIENE, MEDICINE, AND POPULATION

2. OLD DISEASES IN THE THIRD WORLD AND NEW ACHIEVEMENTS IN MEDICINE

1. Jonas Salk (1914–1995) in his research laboratory. The polio vaccine this American biochemist prepared was based on defunct viruses. The first vaccination campaign took place in the United States in 1954 and reduced the incidence of polio by 80%. The microbiologist Albert Sabin (1906–1993) used a weakened living virus as the base of his vaccine. Sabin's vaccine was developed at the same time as Salk's, but because it was easier to administer it soon replaced Salk's throughout the world. (Photograph by FPG Intl/Marka.)

In industrialized countries, the older epidemic diseases, like the bubonic plague and leprosy, are a thing of the past. In these areas of the world, in fact, hygiene led to the control of cholera, of typhus, and, with new pharmaceuticals, of tuberculosis. Diphtheria was defeated by vaccination, as was poliomyelitis (polio) through the research of Salk and Sabin in the 1950s. The final decade of the 20th century saw the first cases of genetic engineering on human beings, with the substitution of new genes for genes responsible for disease. Infectious diseases registered a decline as a cause of death in the 1990s, even though the statistics include the populations of the whole planet.

It is important to note, however, that even at the end of the 20th century 45% of the world's population was at risk for malaria. This percentage will probably increase due to the effects of global warming. The same is true of cholera, still present in South America and Central America—where hemorrhagic fever continues to claim victims—and in central Africa, India, China, and central Asia. Even the plague has not been defeated; it can still be found in India, Mongolia, Congo, and Tanzania. Sulpha drugs and antibiotics have limited the diseases' lethality, but, as occurred in Vietnam and Cambodia, events like wars can renew their virulence. Diphtheria is still present in the countries of the East, where a reduction in vaccination has recently been responsible for the return of tuberculosis.

Aggressive vaccination campaigns seemed to be the way to eradicate the diseases that continued to claim victims in the southern half of the world. In the 1970s, the World Health Organization, with the help of many industrial countries and the collaboration of UNICEF (the United Nations International Children's Emergency Fund, now officially the United Nations Children's Fund) promoted a program aimed at wiping out the diseases thought primarily responsible for infant mortality. The campaign against neonatal tetanus prevented some 700,000 deaths in the 1980s, but the disease still exists. Twenty percent of the victims of tetanus can be found in countries undergoing local wars.

3. A diagnostic examination performed with a new technology: the measurement of radioactive isotopes. In use since 1950, this method has made possible many crucial medical operations, such as explorations of thyroid functioning, studies of anemia, and the testing of the ability to synthesize hemoglobin.

2. Early in the 1950s, the photographer Eugene Smith did a spread on Albert Schweitzer for the magazine Life. Schweitzer, a German medical missionary, had abandoned his academic career and dedicated his entire life to a hospital he had founded in Lambaréné, then in French Equatorial Africa. He later expanded the hospital to include a leper colony. (Illustration by Giuliana Panzeri.)

4. *An automobile-body production line. The lack of specialization, the monotony, and the tedium symbolized by the production line were the cause of violent worker disputes at the end of the 1960s. These led the factories to abandon the production systems established by Ford.*

5. *Automobiles being wrecked. "Car cemeteries" appeared as symbols of an age of waste. Notwithstanding the notable improvements in the field of recycling, the tension between the demands of the environment and those of the economy remains unresolved.*

It is in this field that the automobile, the most successful industrial product of the 20th century, faced disquieting outcomes, besieging cities and causing (sub)urban sprawl. The automobile probably is the principal cause of environmental pollution and the direct and indirect cause of a kind of silent mass destruction; still, the "horseless carriage" continues its reign, always winning new converts.

TRANSPORTATION
7. MAN AND GOODS ON FOUR WHEELS

1. A "country" version of Ford's Model T, ca. 1925. Although born in Europe, the automobile spread rapidly in the United States. Beginning in the years preceding World War I, the automobile helped to draw the inhabitants of the vast American countryside out of isolation.

Vehicles powered by steam motors, like the one developed by the French military engineer Nicolas-Joseph Cugnot in 1770, did not become popular because their usefulness was limited, especially when they were compared to the railroad. By the end of the 19th century, the internal-combustion engine, invented by the Belgian inventor Jean-Joseph-Étienne Lenoir in 1860, marked the beginning of the manufacture of automobiles for an ever-expanding public. The names of the first manufacturers, Karl Benz and Gottlieb Daimler in Germany and Armand Peugeot in France, still carry importance in the automobile industry.

At first, the new vehicles were perceived as "horseless carriages." This title was due in part to the fact that, predominantly in Europe, vehicles were reserved to a limited class of wealthy people. The first automobiles tended to reproduce, in their interior designs and their exterior shape, the aristocratic carriages of the 19th century. Still not abandoning refined and often ostentatious styles, as in the monumental American automobiles, the bodies were quickly adapted to methods of mass production. Starting with Ford's Model T in 1914 and the German Volkswagen, the "people's car," developed between the World War I and World War II, the automobile as we know it today came into common and daily use; and it assumed a significance beyond its basic function. Obtaining a driver's license took on the value of rite of passage to adulthood, and losing one was seen as a sign of social exclusion.

Due to its speed, the automobile was considered the means to eliminate distance, and therefore space and time. Even in the first decades of the 20th century, it symbolized individual freedom; it restored the old mode of traveling that the train's set schedule and obligatory stops seemed to have erased.

Parallel to the image of the automobile was that of the large truck. It became ever more powerful and capable of taking the place of the train as the means of transporting cargo over long distances. Like the racecar drive, the truck driver, master of the road and of solitude, has become a character in the imagination of our times.

2. "Free road to the Marshall Plan." This German poster from 1949 propagandizes American aid to European countries during the reconstruction period that followed World War II. The power and the size of the truck symbolize the driving force of the European Recovery Program (ERP).
3. The speed and the fusion of man and machine are illustrated in this Italian poster designed by Plinio Codognato in 1923, slightly more than 20 years after the first manufacture of the Fiat automobile in Turin, Italy.

3. The remote control, by now an essential item for billions of people who watch television every day, has accentuated the compression that characterizes "television time" as opposed to "natural time." The elimination of dead time, implicit in the language of television, is completely realized in the continual shifting from one channel to another and the ability to watch more than one program at a time.
4. Maurizio Cattelan, Mattia (Madness). Incorporating refuse used artistically, "trash art" finds its raw materials and its inspiration in the thrown-away—including those who live on the fringes, the homeless, who are excluded from the consumer society.

5. A view of urban destruction in Cochabamba, Bolivia. Even in the Third World, rapid commercialization consumed entire urban neighborhoods. Large construction companies seem to have ignored local history and meaning and appreciated only economic value. (Photograph by C. Lavayén.)
6. Costantino Nimola, Clashes in Chicago, 1969. The student riots and the clashes with police characterized both the end of the 1960s and most of the following decade. Incited by the opposition to the Vietnam War in the United States, these confrontations often tied themselves to the workers' protest movements in European countries. (Photograph from a private collection.)

In addition to the knowledge of these market mechanisms, behaviors motivated by basic frugality—not widespread, but significant—were inspired by the cultures of Asia. For example, the Buddhist master Thich Nhat Hanh spoke of awareness of the suffering caused by misguided consumption and the wisdom of avoiding toxins in both foods and other consumables, such as certain television programs. Western-rooted motivations, on the other hand, stemmed from the growing inequality of consumption between the northern and southern parts of the world. More recently, these motivations have been rooted in the certainty that an expansion of Western consumption over the entire planet, inconceivable politically, is not ecologically sustainable.

Vertical section showing the structure of an underground metropolitan railroad to be used for rapid mass transit within large cities and between cities and their suburbs. The first Underground, or subway, opened in London in 1863.

INDEX

Each entry is followed by the number(s) of the chapter(s) in which it appears.

Africa 2, 9, 10
agrarian question 1
Alamogordo, New Mexico 5
Anders, Günther 11
anemia 2
Angola 10
anti-colonial movements: introduction, 10
antibiotics 2
Apennines 1
appliances 3
architecture 3
Asia 2, 9, 10
aspirin 2
assembly line 4
atom 5
atomic bomb 5
atomic physics 5
automobile: introduction, 3, 6, 7, 11
automobile industry 6, 7

balance of power: introduction, 5
Bandung, Java, Indonesia 10
Batista, Fulgencio 10
Benz, Karl 7
bills 4
Black, Duncan 3
Bolivia 11
Brasil 6
bricolage 3
Burma 10
buying power 3,11

Calcutta, India 9
Cambodia 2
Camus, Albert 9
Canada 4
Canary Islands 10
Cape Verde 10
capitalism: introduction, 4, 9, 11
capitalist economy 4, 6, 9
Cassari, Italy 1
Castro, Fidel 10
cathode tube 8
Cattelan, Maurizio 11
cattle 1
César 7
"channel surfing" 9
chemical fertilizers 1
Chernobyl 5
Chicago 5, 9
China 1, 2, 5, 6, 9, 10
chloroflourocarbons 3
cholera 2
cities 3, 6, 7, 8, 9
classes, social strata 3, 4, 8, 10
clothing 11
coal 5
Cochabamba, Bolivia 11
Codognato, Plinio 7
Cold War: introduction, 5, 10
colonization 9
commercial centers (shopping malls) 11
communication 4, 8
commuters, commuting 4
computer science 6
Congo 2
consumption, consumerism: introduction, 3, 6, 9, 10. 11
contemporary art 3, 7, 11
Crick, Francis 2
Cuba 10
Cuban revolution 10
Cugnot, Joesph 7
cyclotron 5

Daimler, Gottlieb 7
Decker, Alonzo 3

decolonization 10
design 3, 7
development: introduction, 9, 11
Diop, Alioune 10
diphtheria 2
dishwasher 3
division of labor (Taylor-Ford) 4
DNA (deoxyribonucleic acid) 2
Dominican Republic: introduction
Duisburg, German 9

ecological and feminism movement 10
economic expansion 4
Einstein, Albert 5
electric drill 3
electronic imaging 8
electronic games 3
electronics 6
emerging countries 9
emigrants 1
energy crisis 5
energy: introduction, 1, 5
Eniwetok 5
entrepreneurs, enterprises 1, 4, 8
environment: introduction, 6, 8, 9
ERP (European Recovery Program) 7
Europe 1, 3, 4, 6, 7, 8
European Community 1

family 3, 4, 10, 11
farm machinery 1
farmers 1
Fermi, Enrico 5
Ferrari automobiles 6
Fiat (Fabbrica Italiana Automobili Torino) 7
financing, of the economy 4, 9
Ford automobiles 6, 7
former colonial nations 10
France 5, 7
Franco, Francisco 10
free time 3
free trade, *laissez-faire* 1
French Equatorial Africa 2
fruit and vegetable farming 1
fundamental rights of man 10

gene therapy 2
General Motors 6
genetics 2
Germany 6, 7, 9. 10
Ghana 11
Gibuti 10
Gide, André 10
global warming 2
Goodyear, Charles 6
Great Britain 5, 10
Gross National Product (GNP) 9
Guinea-Bissau 10

Hanh, Nhat 11
harvester-thresher 1
Havana, Cuba 10
heating 5
hemoglobin 2
hemorrhagic fever 2
Hiroshima 5
homeless 11
Hopper, Edward 8
house, residence, apartment 3
human relations 4
hydrogenolysic imbalance 1
hygiene 2, 3

Illinois, USA 9
incentives, economic 6
India 2, 6, 9. 10

30